Module 1:

Introduction to Environmental Economics

Environmental Economics: Meaning, Definition and Scope

1.0 Introduction

1.1 Definition of Environmental Economics

1.2 Need for Special Course in Environmental Economics

1.3 The Scope of Environmental Economics

1.0 INTRODUCTION

In the early tradition of economic thought, laid down by the classical and neo-classical economists, economics was conceived as the study of the allocation of scarce resources among alternative, competing ends, in the quest to satisfy human wants. The classical and neo-classical economists underestimated the environmental issues of production and consumption, since they considered these as social issues. Consequently, the impacts of production and consumption on the natural environment were not explicitly brought into the mainstream of economic theory. The environment was taken as given and can last ad infinitum.

However, in the early 1970s, the reality of economic thought brought to a halt the perception of economics merely as a science of production and distribution. In the new thinking, economics is no longer perceived as the science of production and distribution which neglects the environmental repercussions of economic activities. This means that economics as a subject cannot exist in isolation. It must take into consideration the effects of resource use in production and distribution on the natural environment which supplies these resource inputs. Thus, any study on the economic content of production, distribution, consumption, development etc cannot be completed without touching upon the environmental aspects like pollution of the environment (e.g water and air pollution), environmental damage, environmental resource exhaustion and depletion, global warming, biodiversity, social externalities and the likes.

In order to account for the environmental impact of economic activities, a new field of study called environmental economics thus emerged. The call for an evaluation of environmental impact of economic activities necessitates the regulation of these activities for environmental sustainability. This regulation therefore brings into force the issue of environmental policy, which nearly all countries all over the world now spend resources to formulate at national and international levels.

In order to account for the environmental impact of economic activities, a new field of study called environmental economics thus emerged. Although, this branch of economics can be traced to the 1950s and the 1960s, with important contributions from the think tank on resources for the future, the field really took off from the 1970s and has been booming ever since. Environmental economics can therefore be seen as an applied part of economics which deals with the entrepreneurship between economic activities and the environment and studies the way and means by which the former is not impeded nor the latter impaired.

1.1 DEFINITION OF ENVIRONMENTAL ECONOMICS

Environmental economics is concerned with the analysis of the impact of the economy on the environment, the significance of the environment to the economy and appropriate way of regulating economic activity so that balance is achieved among environmental, economic and other social objectives.

It should be noted that environmental study requires a synthesis of the various branchesof knowledge like science, economics, philosophy, ethics, anthropology etc. Therefore, the studyof the environment may be approached from different perspectives and environmental economics may borrow from such perspectives in its analysis.

1.2 NEED FOR A SPECIAL COURSE IN ENVIRONMENT ECONOMICS?

Environmental and natural resources should be allocated in the same way we want to allocate all resources. That is, efficiently and equitably. In this sense, there is no theoretical need to have a separate course for natural and environmental resources. Then why do we do courses in environmental and natural resource economics? Economic analysis of the environment is challenging and important precisely because environmental value is not always conveniently revealed in a market, and thus is subject to inappropriate use. Indeed, lots of people are particularly worried about the allocation of environmental and natural resources. And, a lot of people think they are being misallocated. This has generated some concerns which gave credence to the separate treatment of environmental and natural resources. Some of these concerns are:

1. Natural resources are finite and stocks are dwindling, so may be the shit has to hit the fan sometime soon.

2. Increasing concern about non-catastrophic pollution. This result from increasing information about the effect of pollutant on health etc.

3. Concern about catastrophic pollution; that is, pollution that could lead to a widespread eco disaster such as green-house effect, loss of the ozone layer and radiation from bombs or leaks.

4. Concern over the preservation of natural environments: Wilderness areas, National Parks, the rain forests, the Arctic, etc, hence Green Association such as Sierra Club, Friends of the Earth, National Wildlife Federation etc.

5. Concern over preservation of animal species. The possibility of extinction is a growing concern for many species; hence we have conservation groups such as Greenpeace, World Wildlife Fund, Trout Unlimited, Ducks Unlimited etc.

6. Consequently, economist is increasingly being called upon to help guide the process of environmental management and to provide measurable criteria by which environmental policies can be evaluated. Hence, the need for a study on environmental economics.

1.3 THE SCOPE OF ENVIRONMENTAL ECONOMICS

As a sub discipline of economics, environmental economics originated early years of the so-called environmental movement. However, despite its brief history, over the past three decades it has become one of the fastest-growing fields of study in economics. The growing popularity of this field of inquiry parallels the increasing awareness of the interconnectedness between the economy and the environment-more specifically, the increasing recognition of the significant roles that nature plays in the economic process as well as in the formation of economic value.

The nature and scope of the issues addressed in environmental economics are quite varied and all-encompassing. Below is a list of some of the major topics addressed in this field of study.

The causes of environmental degradation.

The need to re-establish the disciplinary ties between ecology and economics.

The difficulties associated with assigning ownership right to environmental resources.

- The trade-off between environmental degradation and economic goods and services.
- The ineffectiveness of the market, if left alone, in allocating environmental resources.
- Assessing the monetary value of environmental damage.
- Public policy instruments that can be used to slow, halt and reverse the deterioration of environmental resources and/or the overexploitation of renewable and non-renewable resources.
- The macroeconomic effects of environmental regulation and other resource conservative policies.
- The extent to which technology can be used as a means of ameliorating environmental degradation or resource scarcity, in general- that is, limits to technology.
- Environmental problems that transcend national boundaries, and thus require international cooperation for their resolution.
- The limits of economic growth.
- The extent to which past experience can be used to predict the future events that are characterized by considerable economic, technological and ecological uncertainties.
- Ethical and moral imperatives for environmental resource conservation- concern for the welfare of future generations.
- The interrelationship among population, poverty and environmental degradation in the developing countries of the world.
- The necessity and viability of sustainable development.

This list by no means exhausts the issues that can be addressed in environmental economics. However, the issues in the list do provide important clues to some of the fundamental ways in which the study of environmental economics is different from other subdisciplines in economics.

First, the ultimate limits to environmental resource availability are imposed by nature. That is, their origin, interactions and reproductive capacity are largely governed by nature.

Second, most of these resources have no readily available markets: for example, clean air, ozone, the genetic pool of a species, etc.

Third, no serious study of environmental economics can be entirely descriptive. Normative issues such as intergenerational fairness and the distribution of resources between the poor and rich nations are very important.

Fourth, uncertainties are unavoidable consideration in any serious study of environmental and natural resource issues. These uncertainties may take several forms, such as prices, irreversible environmental damage, or unexpected and sudden species extinction. Such is the nature of the subject matter of environment economics.

Module 1

Structure of the Environment

2.0 – Introduction

2.1 – Air Environment

2.2 – Water Environment (Hydrosphere)

2.3 – Land Environment   (Geosphere)

2.4 – Biological Environment (Biosphere)

    2.4.1 – Biosphere and Its Dependent Links to Other Spheres

    2.4.2 - Biosphere and Its Impact Links to Other Spheres

2.5 - Astrosphere

2.6 – Social Environment

2.0 Introduction

The structure of the environment can be described in terms of its interdependent components, namely air, water, land living system and social structure. However, there are other two aspects that impact on the natural environment, these are: (a) Anthrosphere (b) Social environment

2.1: The Air Environment (Atmosphere): This includes air and the atmosphere. The atmosphere reaches over 550 kilometers from the surface of the earth. Some significance of the atmosphere can outline as follows:

a). The atmosphere contains life-supporting gasses for plants and animals. The presence of carbon dioxide and oxygen is essential for the coexistence of the basic living systems of plants and animals.

b). Nitrogen gas present in the atmosphere is an essential component of plants, animals and microbes.

c). The atmosphere is a shield that protects life on earth from the hostile conditions from the outer spheres.

d). The atmosphere offers mankind a field of study called meteorology that deals with weather and climate change which have profound influence on other spheres of the environment.

e). A major layer in the atmosphere is the ozone layer which protects human, plants and otheranimals from hazardous radiation from the sun.

2.2 Water Environment (Hydrosphere): The water environment constitutes three-quarters of the planet Earth. All plants and animals (including humans) on earth depend on the availability and quality of water for survival. The importance of water cannot be overemphasized, it includes:

1. All forms of life depend on water for survival, for instance:

    a). Water energizes food particles, which are able to supply the body with energy during digestion

    b). Water increases the rate of absorption of essential substances in food

    c). Water increases the efficiency of red blood cell in collecting oxygen in the lungs

    d). Water carries every nutrient, mineral, vitamin, protein, hormone and chemical messenger in the body to its destination

    e). Water is essential to the cleaning process of the body (cleanses internal organs andeliminates toxins from the bloodstream) etc

2. Water is an essential requirement for all agricultural, aqua cultural and industrial activities ofman.

In the light of the above we can see that water environment is needed both from the maintenanceand sustenance of plant and animal (including humans) life, and the economy.

2.3 Land Environment

Land environment comprises of geospheric components like rocks, soils, and other associated constituents. The importance of land resource lies in the fact that it is the source of all our vital requirements [such as building materials, minerals (rocks), fuels, soil, wilderness areas etc] that are needed for living.

2.4 Biological Environment (Biosphere)

This consists of all aspects of living systems – microorganisms, macro plants, and macro animals. It should be noted that humans are a part of the biological environment and that there is a great diversity and uniqueness of life forms inhabiting the biosphere. The biological environment is the global ecological system integrating all living beings and their interactions among each other and between living things (especially humans) and the other components of the natural environment. But for the presence and interactive nature of living systems, the energy and material world would come to a grinding halt and our planet would lose its unique position among celestial bodies.

There are two ways in which the biological environment is linked to other components of the natural environment. The first is the dependence of the entire biosphere on itself and other natural environment components. The second is the impact link of the biological environment (the human activity sub-component) to itself and other components of the natural environment.

2.4.1 Biosphere and its Dependent Links to the other Spheres

For sustenance of life the biosphere depends on itself and other spheres of the environment. Some of the ways the biological environment depends on other components of the natural environment are highlighted below:

a. <u>Dependence on atmosphere</u>: Life processes involve a vast number of chemical reactions, some of which either extract or emit gases from and to the atmosphere which acts as the store house for these gases. For instance, photosynthesis absorbs CO_2 from and produces oxygen to the atmosphere.

Hence the atmospheric component of the environment should not be damaged so that both plant and animal (including human) life can continue to receive the support of the atmosphere.

b. <u>Dependence on Hydrosphere</u>: The biological environment requires water to survive because water is essential for all living organisms in the biosphere and has played a key role in the evolution and sustenance of life on our planet. Water is also important in transporting soluble nutrients (phosphate and nitrate) that are needed for plant growth, and for transporting the waste products of life's chemical reactions. This means that the water environment component of the natural environment should not be overused or misused for the satisfaction of present needs since water is still needed to sustain life after our present generation.

c. <u>Dependence on Geosphere</u>: The land (geosphere) and biological (biosphere) components of the natural environment are intimately connected through soil which consists of a mixture of air, mineral matter, organic matter, and water. Indeed, soil is important for both plant and animal life. Mineral's exploitation and mining activities for human survival also depend on the land. One pertinent environment issue is how to use land sustainably.

2.4.2 Biosphere and its links to other spheres (components) of the environment: The biological environment (biosphere) is of great significance to us for two principal reasons:

(i) Humans are part of the biological environment and so should naturally be concerned about it.

(ii) The changes that occur in the biosphere itself and in other spheres (components) of natural environment originate from human activities.

The point raised in (ii) emphasize that an important part of the subject matter of environmental economics is the study of human activity impact on all the components of the natural environment (please see the definition of EE and identify the other two important aspects of EE as defined).

In order to impact on the four basic components of the natural environment, humans created another sphere of the environment called ANTHROSPHERE.

2.5 The Anthrosphere is that part of the environment that is made or modified by humans for use in human activities. It is also called the techno sphere. As human technology becomes more evolved, so do the impacts of human activities on the natural environment.

Below are some of the ways that the anthrosphere has impacted on other components of the natural environment:

a. Atmosphere: Industrial and agricultural activities have changed the composition of the atmosphere (the air environment). For example, we have increased the concentration of carbon dioxide in the atmosphere by not less than 26% and doubled the concentration of methane gas. The productions of chlorofluorocarbons are depleting the earth's ozone layer, our natural defense against ultraviolent radiation. Man has also affected the quality of air (e.g., smog), especially in urban areas which result in respiratory problems

b. <u>Hydrosphere:</u> Humans have impacted the water environment by withdrawing large amounts of groundwater for agriculture and contaminating rivers, lakes, groundwater and oceans by organic and industrial wastes.

c. <u>Biosphere:</u> Humans have already altered the biological environment, of which they are a part, through economic activities. A prime example is the slash and burn agricultural practice in the tropics where rainforest is cut and burned and the land is converted to pasture.

d. <u>Geosphere:</u> Minerals and energy resources (coal, crude oil, coke, etc) from the land environment have fueled the industrial revolution that has permitted the human specie to increase so prodigiously in number. For example, the exploitation of fossil fuels has increased our standard of living but an unintended consequence of this action may be climate change and global warming.

The ultimate goal of environmental economic is to minimize the flux of pollutants and/or toxic substances across the interface between the astrosphere and other spheres while maintaining a functional technological society.

2.6 SOCIAL ENVIRONMENT

The social environment refers to the combined structure involving human to human interactions. The social structure of a human community is made up of;

a) The constraint environment which includes the physical, biological and chemical constraints to which human activity is subject.

b) Infrastructure, defined here to include;
   i. Mode of production which deals with the technology and the practices employed in expanding or limiting production and the ensuing techno-environmental relationships.
   ii. Mode of reproduction which deals with the technology and practices employed for expanding, limiting and maintaining population size.
c) Structure which refers to relationships or bonds between group of individuals in a society and include;
   i. Primary group structure: such family members, community, friendship networks, voluntary organizations who interact in an intimate basis and enforce socially acceptance values. The primary group can be a potent instrument in disseminating environmental education. They perform such functions as regulating reproduction, basic production, socialization, education, and enforcing domestic discipline.
   ii. Secondary group structure: A small or large group of individuals which members interact without emotional commitment to one another. They may include government, parties, factions, military and police, businesses and industries, media, non-governmental organizations, professional association, labour union etc. These organizations are coordinated through bureaucracies. They perform many functions such as regulating production, reproduction, socialization, education and enforcing social discipline. They are a potent source of environmental policy formulation and implementation.
   iii. Superstructure which is made up of the cultural superstructure (art, music, dance, literature, rituals, sports, games, science etc) and mental superstructure (that is conscious

and unconscious motives for human behavior (values, emotions, traditions). The superstructure influences the way human interact with the environment.

The Human Economy and the Natural EnvironmentIntroduction:

The natural environment could be defined as physical, chemical, and biological surroundings thatcomprise the Earth's endowment of life-support systems. It includes air, water, land (the Earth's crust), living system, and even radiation from the sun. The natural environment is the supplier of all natural resources like arable land, wilderness areas, mineral fuels, nonfuel minerals, watersheds, the ability of the natural environment to degrade waste and absorb ultraviolet light from the sun, etc. These resources can be regarded as natural capital from which human economy draws upon.

The human economy consists of all the production and consumption activities of human beings. Such production and consumption activities can occur on land, beneath the Earth's crust, in seas, in the atmosphere and outer space. One main characteristic of the human economy is thatit produces both useful commodities (goods and services) and non-useful materials called wastes.

The interlinkages between the human economy and natural environmental system cannot be overemphasized. Humans cannot exist in isolation of the natural system because all economic (even non-economic) activities of man are interconnected with the various other living and non- living endowments of nature. Similarly, the complex physical, chemical, and biotic factors of the natural system act on man and ultimately determine human form and survival.

The economy depends upon natural capital and reconfigures that natural capital to produce products to satisfy human wants. One characteristics of the human economy is that it produces intended useful product and some by-products. The by-products of an economic reconfiguration process are referred to as "waste" since the production of these by-products is not the central object of reconfiguration. The production of "waste" from any reconfiguration of natural capital is a necessary consequence of the first law of thermodynamics. The law says that matter is neither created nor destroyed in chemical reactions, although it may be transformed from one form to another. Any wastes from production and consumption activities are absorbed by the natural environment.

The ongoing reconfiguration of natural capital in the natural system means that the system also acts as a recycling process, this recycling role of the natural system means that it is possible for the economy to continually reconfigure natural capital for human purposes, and for that reconfiguration to be continually reversed through the natural system. Unfortunately, there are limits to the rate to which materials can be recycled through the natural system in this way. This is to say that the natural system has assimilative capacity. This is sometimes called a "throughput constraint".

If human activity produces more "wastes" than can be naturally recycled, then, there is a flow of waste in excess of assimilative capacity. This can cause the natural system to change in possibly drastic and irreversible ways. It should be noted however that drastic and irreversible change in the natural system is not due solely to human activity. For example, volcanic eruptions produce vast quantities of pollution. The relationship between human economy and natural environment can be explained in the form of a "Material Balance Models" developed by Alen Kneeze and R. V. Ayres. The material balance model is based on the first and second law of thermodynamics. These models consider the total economic process as a physical balanced flow between inputs and output.

Inputs are bestowed with physical property of energy which is received from the sun. the resulting output from inputs carries the same level of energy. Similar to this, there are wastes resulting from consumption activities. Materials and energy are drawn from the environment, which are used for production and consumption activities and returned to environment as waste. So far as this balance is maintained, there are no environmental issues. The material balance model of the economy is given in figure 1.1

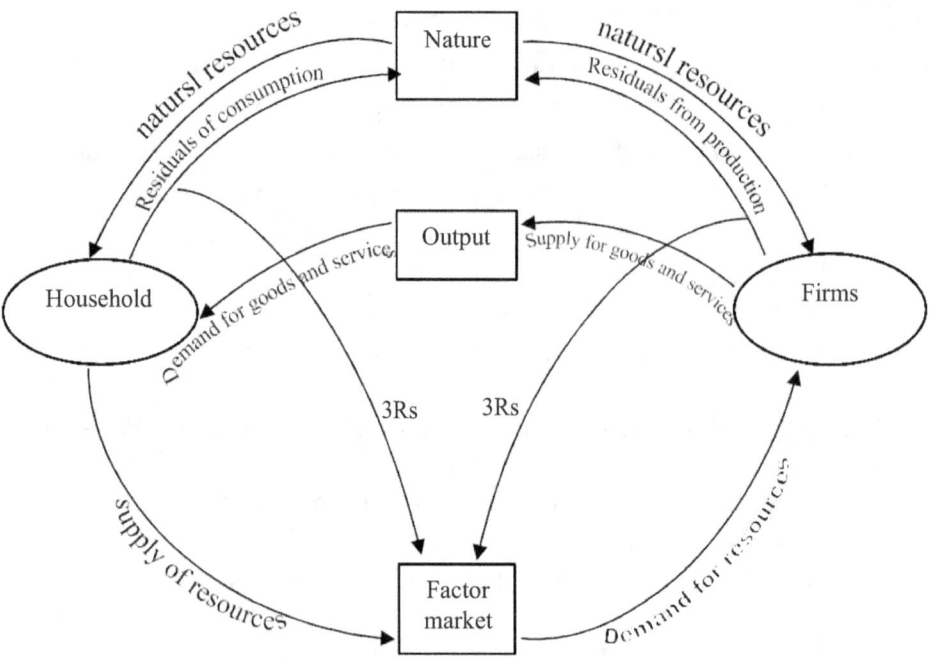

Fig 1.1 the material balance model: independence of economics and environment

The figure 1.1 shows that environment is the supplier of all forms of resources like renewable and non-renewable, and it is also acting as a sinking for cleaning up of waste. Households and firms are connected to environment, and they are connected too. Households and firms depend on the nature of resources. Both households and firms send out residual consumption and production respectively to nature. As mentioned earlier nature has the power to assimilate all forms of waste. But this power is conditional. So long as earth is not being disturbed by the excess amount of waste, the earth can clean up natural waste. When the earth fails to respond to 3Rs, the symptoms of environmental damage appear. Thus, there is a rhythm in the use and reuse of resources for men by men: earth cannot respond properly to man-made or artificial waste. Man-made wastes are piling up around us, and therefore, the extent of damage to the environment has been on the rise.

All the waste that has been sent out cannot be cleaned up by the sink earth. As long as earth can discharge this function of cleaning up of pollution due to waste, there would not be any environmental issue. But earth has reached at the saturation point of this process, and it is helpless in cleaning up of several types of wastes resulting in major environmental issues in the world over.

The impact of the transformation of material inputs and energy into output is subject to several changes in the biosphere. The process of transformation is better explained with the help of the laws of thermodynamics. The first two laws of thermodynamics are worth mentioning in the context. The first law of thermodynamics, which is often referred to as the law of conservation of matter and energy says that energy, like matter, can neither be created nor destroyed, but at the same time the forms of energy can be transformed. The law stresses that the total amount of energy created through production and consumption activities must be equal to the total sum of initial energy extracted from the nature. Therefore, the first law of thermodynamics implies the accounting identities of material balance model.

The second law of thermodynamics is known as the law of entropy. Entropy is usually considered as the measure of unavailability of the benefits of energy or simply wastes. When one form of energy is transformed into another (say for example, when the thermal energy of coal is converted into electrical energy) there is waste of energy, and the volume of waste depends upon the technological process. Entropy will be low, when materials and energy are highly structured and organized. When a piece of coal is kept idle, there is low entropy, but when it is burnt up, the same piece of coal is subject to high entropy, since heat and carbon dioxide are dissipated, but sometimes unavailable for use.

Thus, the second law says that as long as there is utilization of material inputs and energy for production and consumption activities, the level of entropy will be high. Economic activity helps to convert low entropy resources and energy into high entropy waste that is, resources into wastes. Economic activities cannot be stopped on account of high entropy, but at the same time, through recycling and waste management, it is possible to bring into the economic system, low entropy value. Use of natural resources, but at the same time with minimal waste or damage to the environment is considered as the key theme of sustainable development. It is a form of development path that is ready to meet the needs (not greed) of the present generation, at the same time without compromising the needs of posterity. A detailed discussion of sustainable development is included as a separate chapter of this book.

We must know that the environment discharges the following economic functions:

1. The environment is the supplier of all forms of resources.
2. The wastes are cleaned up by the environment.
3. The environment maintains genetic diversity and stabilizes the ecosystem.

The above-mentioned functions of the environment are interlinked. In the name of economic activity, the environmental resources are transformed into economic goods [converting low entropy resources into high entropy ones]. In this process of transformation, wastes are created. Resources are also getting depleted due to the overuse. When environment is disturbed by the overuse and the huge number of wastes, it cannot discharge the third function that is, maintaining genetic diversity and stabilization of ecosystems. It further affects the life and existence of flora and fauna. Therefore, an integrated approach to the study of economy, ecology, and environment is essential, as all these are closely interlinked.

The Neoclassical Perspective of the Human Economy and Natural Environment:

The mainstream Economists have a particular conception of the natural environment, including how it should be managed. The conception emanates from the Classical and neoclassical dominant approach to economic analysis since about 1870s.

The Neoclassical worldview of environmental economic relationship is anthropocentric. This means that the humans are treated on pre-eminent in the natural environment. Consequently, the human economy is rated above natural environment and humans are regarded as the universe's most important entity. The natural environment therefore exists to serve the human economy and environmental resources have no intrinsic value. Sequel to this worldview, the economy is assumed to depend on the environment for three distinctive purposes:

a). the extraction of nonrenewable resources (such as iron ore, fossil fuels, etc.)and the harvest of renewable resources (such as fish of various species, agricultural products, forest products, etc.) to be used as factors of production;(Arrow 1)

b). the disposal and assimilation of wastes ;(Arrow 2) and

c). the consumption of environmental amenities (such as bird watching, canoeing, hiking national park trails, observing a morning sunrise or an evening sunset, etc.) (Arrow 3).

Thus, broadly viewed the economy is assumed to be completely dependent on the natural environment for raw materials, the disposal of waste material and amenities

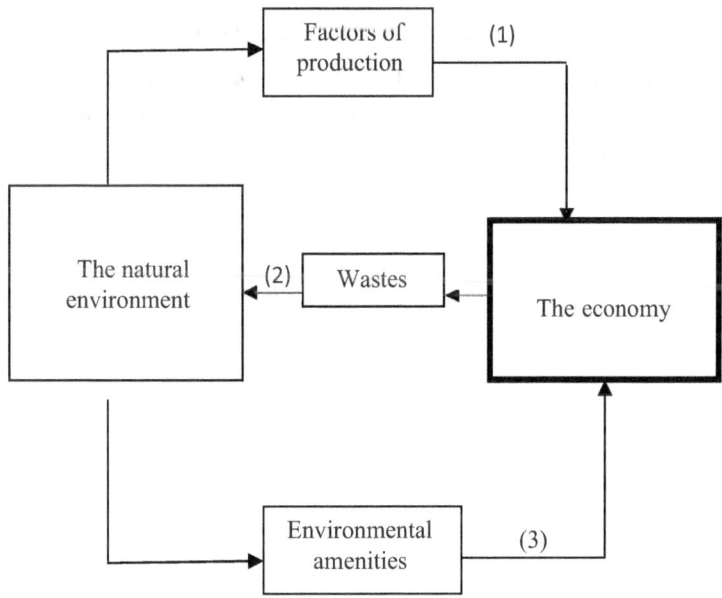

Figure 1.1 A schematic view of how the human economy depends on the natural environment for factors of production, disposal of waste and consumption of amenities.

Furthermore, since the earth is 'finite' there exists a theoretical upper limit for resource extraction and harvest and disposal of waste into the natural environment. Thequalities of environmental amenities and the maintenance of life support systems (such as climate regulation and genetic diversity) are also affected adversely in direct proportion to the amount resource extractions and/or harvesting and the disposal or discharge of waste into the natural environment. Thus, as with any other branch of economics, fundamental to the study of environmental economics is the problem of scarcity—the trade-off between economic goods and the preservation of environmental quality. There are some fundamental assumptions that the standard economics approach uses in addressing this subject matter; these are outlined below.

> Environmental (Natural) resources are 'essential' factors of production. A certain minimum number of natural resources is needed to produce goods and services. Environmental resources are of economic concern to the extent that they are scarce.
>
> The economic value of natural resources (including the service of natural ecosystems) is determined by consumers' preferences, and these preferences are best expressed by a freely operating private market system.
>
> Market price can be used as a measure (indicator) of resource scarcity, including the environment.
>
> In both the production and the consumption sectors of an economy, a specific natural resource can always be replaced (partially or fully) by the use of other resources that are either man-made (manufactured) or natural.
>
> Technological advances continually augment the scarcity of natural resources. Nothing is lost in treating the human economy in isolation from the natural ecosystems- the physical, chemical and biological surroundings that humans and other living species depend on as a life support. That is, the natural ecosystem is treated as being outside the human economy and exogenously determined.
>
> Note that to indicate this, in Figure 1.1 the human economy and the natural environment are drawn as two distinctly separate entities.

Clearly, from the above discussions it should be evident that, at the fundamental level, central to the neoclassical economics worldview with respect to the natural environment and its role in the economic process are the following four key issues:

i. the market as a provider of information about resource scarcity;

ii. resource (factor) substitution;

iii. scarcity augmenting technological advance; and

iv. the nature of relationships between the human economy and the natural environment.

The rest of this chapter will address these four issues one at a time.

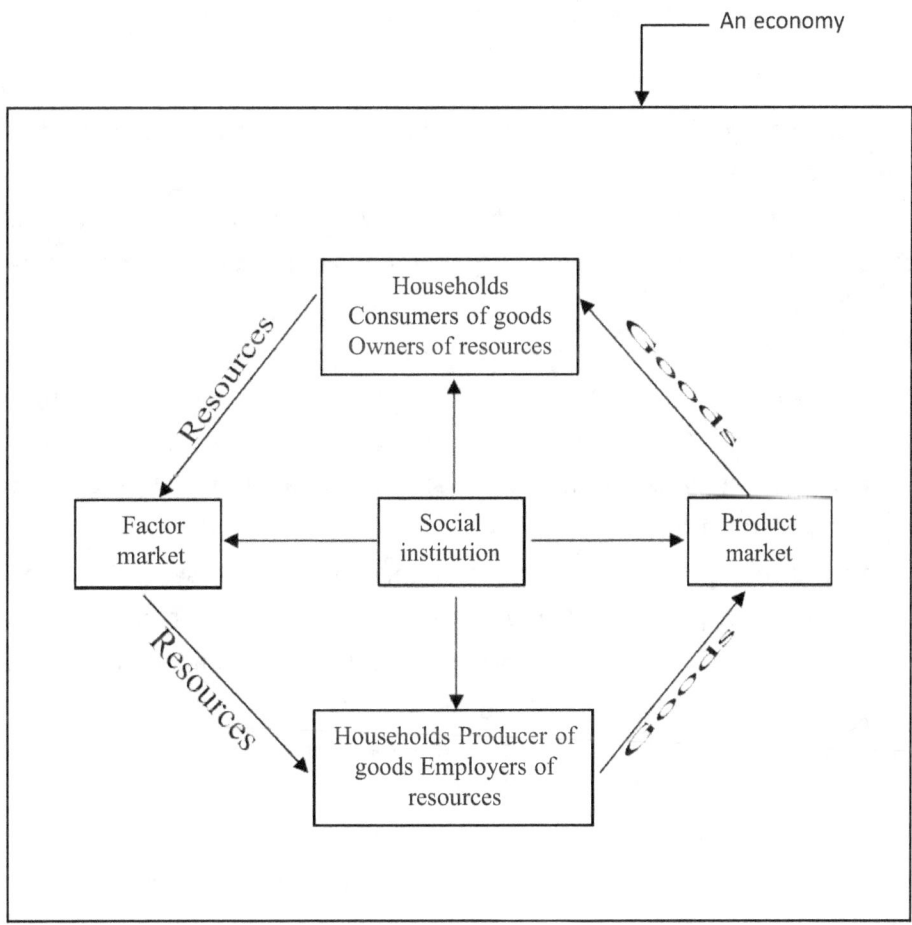

Figure1.5: Circular flow diagram of the economic process. An economy is composed of a flow of *commodities* (in the form of basic resources, goods and services): *social institutions* (primary markets and legal tenders); and *people* (broadly identified as households and firms).

According to the neoclassical worldview, the human economy, as depicted in the abovediagram, is composed of people, flows of commodities (or flows of matter-energy at the fundamental level) and human institutions. The primary focus of the human economic system isnot so much on the conversion of matter-energy that are found in nature to goods and services (i.e., the production process) but generation of utility – an immaterial flux of satisfaction to humans. In this worldview, it appears that the link between the flow of matter-energy in the economic system and the natural environment is very much ignored.

B   An ecological perspective of the Natural Environment and the Human Economy

The ecological perspective of the relative between the natural environment and the human economy is biocentric; this implies that the human economy is not viewed in isolation from the natural ecosystems. The basic objective of the ecological perspective is therefore to establish a clear understanding of the basic principles governing the nature, structure and functions of the biosphere (and by extension, environment resources) and the functional relationship between the biosphere and the human economy.

From a purely ecological perspective, these basic principles and linkage are identified as follows

Environmental resources of the biosphere are finite. Hence, environmental resources are scarce in absolute terms.

In nature, everything is related to everything else. Moreover, survival of the biosphere requires recognition of the mutual interdependences among all the elements that constitute the biosphere.

At a functional level and from a purely physical viewpoint, the biosphere is characterized by a continuous transformation of matter and energy. Furthermore, the transformation of matter and energy are governed by some immutable natural laws. Material recycling is essential for the growth and revitalization of all the subsystems of the biosphere, including the human economy subsystem.

Nothing remains constant in nature. Furthermore, changes in ecosystems do not appear to occur in an absolutely linear and predictable manner. However, measured on a geological time scale, the natural tendency of an ecological community (species of plants, animals and micro-organisms living together) is to progress from simple and unstable relationships (pioneer stage) to a more stable, resilient, diverse, and complex community.

The human economy Is a subsystem of biosphere and it would be dangerously misleading to view natural resources as just factors of production lying outside the confines of the larger system.

The natural tendency of human technology is towards the simplification of the natural system, eventually leading toward less stable, less resilient and less diverse ecological communities

Figure attempts to portray a worldview that is consistent with these principles, and more specifically the ecological (biocentric) perspective of the relationship between the

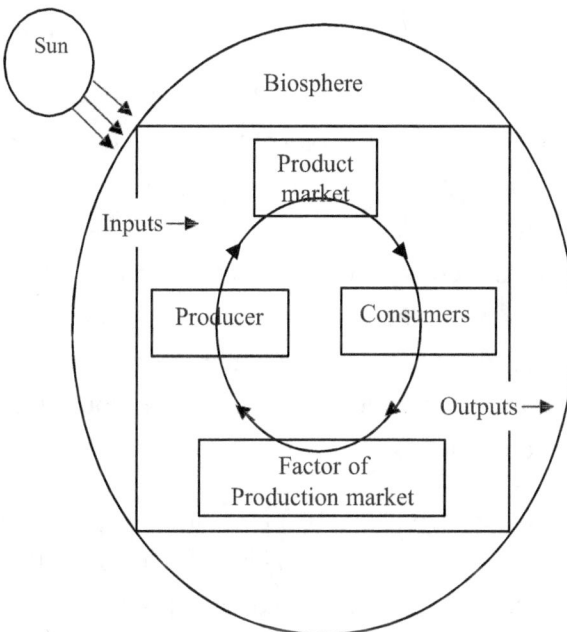

Figure 2.1 Ecologically enlightened economic view. The biosphere is continuously energized by solar power. The human economy (comprising the activities in the inner circle) depends on inputs (throughout) and outputs (disposal of waste) to the biosphere. The biosphere is finite, as indicated by the outer circle.

Biosphere and the human economy. This perspective is biocentric in the sense that it does not explicitly recognize the main output of the economic system – non-material lows of utility (enjoyment). It describes nature and the interactions that occur in nature between living and non- living matter in purely physical (energy and matter) terms.

These features are clearly evident in the following specific aspects of Figure 2.1. First, a clearly demarcated circle, perhaps symbolizing the Earth and its finiteness, represents the biosphere.

Second, by locating it inside the circle, the human economy is perceived as a subsystem of biosphere. The box inside the circle indicates that growth of the economic subsystem is 'bounded' by a non-growing and finite ecological sphere.

Third. Figure 2.1 suggests that the human economy is dependent on the biosphere for its continuous withdrawal (extraction and harvest) of material inputs and as a repository for its waste (outputs)—degraded matter and energy that are the eventual by products of the economic process.

Fourth, the biosphere (and hence the human economy) requires a continuous flow of external energy-- from the sun.

Fifth, while both the human economy and the biosphere are regarded as an "open system" with regard to energy (i.e., both systems require am external source of energy), the biosphere taken in its entirety is regarded as a "closed system" with respect to matter. Note that this is in stark contrast to the way the human economy is depicted in figure 1.5 – the circular-flow diagram discussed in section X of lecture three diagram actually treats the human economy as an 'open system' with regards to both energy and matter. That is, the human economy is continuously dependent on external (outside) sources for input of energy and matter and on external repositories for its outputs.

An ecological worldview as represented in Figure 2 appears to incorporate the principle that the human economy is completely and unambiguously dependent on natural ecological systems for its material needs. Furthermore, the human economy (as a subsystem) cannot outgrow the biosphere. The implication of this is that, as mentioned earlier, the growth of the economic subsystem is 'bounded' by a non-growing and finite ecological sphere. A comprehensive and systematic understanding of the extent to which nature acts as both a source of and a limiting factor on the basic material requirements for the human economy therefore, demands some level of understanding of ecology.

Ecology is a branch of science that systematically studies the relationships between living organisms and the physical and chemical environment in which they live.

MODULE 2: SOME MICROECONOMIC CONCEPTS FOR UNDERSTANDING ENVIRONMENTAL ECONOMICS

1.0 Introduction

In this module we shall be concerned with five concepts in microeconomic theory that aid our understanding of some of the issues in environmental economics. Such concepts will be considered and these are the concepts of scarcity, resources, demand and supply, market, marketfailure, externality and property right.

SCARCITY AND RESOURCES

1.1 The Concept of Scarcity

Definition: A scarce resource is one which, when offered to people at no cost, more would be wanted (demanded) than is available (supplied). Notice that the opposite of a scarce resource is afree resource. At no price, the quantity supplied of a free good for instance exceeds the quantity demanded leading to a surplus.

The definition of scarcity above is further explained graphically as shown below:

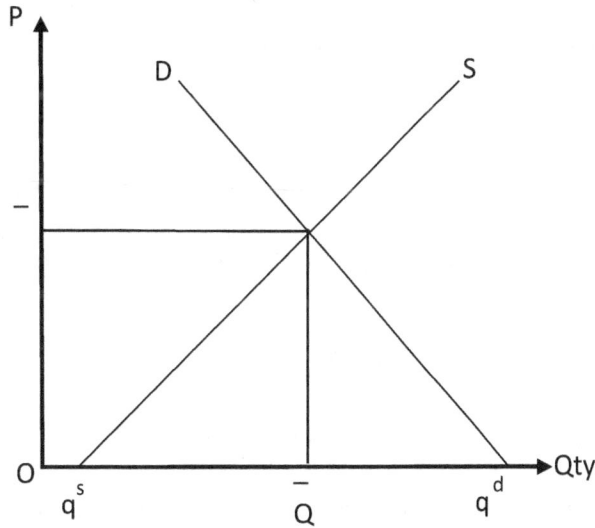

Fig x: demand and supply and market clearing (equilibrium) price for a scarce resource. Notice that at zero price (on the price axis, zero price coincides with the origin), quantity demanded is q^d and quantity supplied is q^s. Notice further that q^d far exceeds q^s, creating a shortage or scarcity.

On the other hand, at zero price, the quantity demanded of a free good is smaller than the quantity supplied, creating a surplus. Consider oxygen which is freely supplied by nature, the availability of oxygen from the ambient air (supply) far exceeds the quantity demanded in a non-polluted environment. Thus, oxygen may be treated as a free good. The demand and supply analysis for a free good is drawn below.

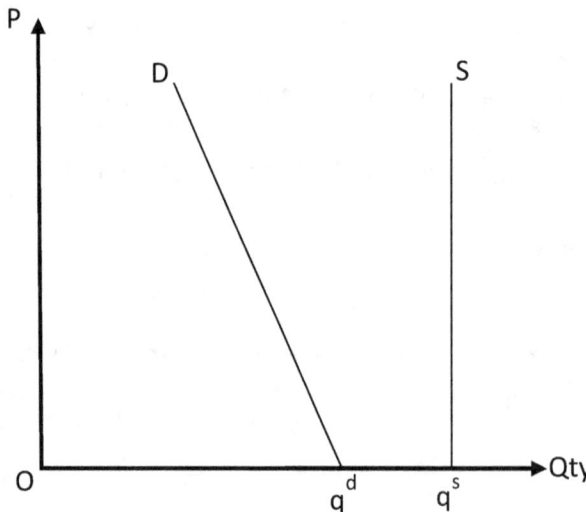

Notice that at zero price (point on the origin), the quantity supplied far exceeds the quantity demanded as q^s is greater than q^d, counting from the origin in economics, the situation of scarcity arises when there is less of an economic good, service or resource than people would like to have if it were free. Scarcity reflects the fact that there are not sufficient resources (inputs) to produce everything that individuals want.

It should be noted that in the absence of scarcity, no difficult choices would need to be made and hence no opportunity or real cost, no prices would need to be attached to anything, andthe study of economics would be rendered entirely unnecessary. Furthermore, as the economist uses the concept, scarcity is present in all societies whether rich or poor in as much as there is a gap between resource need and resource availability.

1.2 Economic Implication of Scarcity

Considering that human wants for goods and services are immense and, worse yet, insatiable in aworld of scarcity, what can be done to maximize the set goods and services that people of a given society can have at a point in time? This question clearly suggests that the significant economic problem involves rationing limited resources to satisfy human wants and, accordingly, has the following four implications:

Choice- the most implication of scarcity is the need to choose. That is, in a world of scarcity, we cannot attain the satisfaction of all our material needs completely. Hence, we need to make choices and set priorities.

Opportunity cost- every choice we make has a cost associated with it; one cannot get more of something without giving up something else. In other words, an economic choice always entails sacrifice or opportunity cost- the highest-valued alternative that must be satisfied to attain something or satisfy a want. In a world of scarcity, "there is no such thing as a free lunch."

Efficiency – in the presence of scarcity, no individual or society can afford to be wasteful or inefficient. The objective is, therefore, to maximize the desired goods and services that can be obtained from a given set of resources. This state of affair is attained when resources are fully utilized (full employment) and used for what they are best suited in terms of production (i.e., there is no misallocation of resources). Furthermore, efficiency implies that the best available technology is being used (McConnell and Bruce 1996). Social institutions- as noted earlier, the essence of scarcity lies in the fact that people's desire or goods and services exceeds society's ability to produce them at appoint in time. In the presence of scarcity, therefore, the allocation and distribution of resources always cause conflicts. To resolve these conflicts in a systematic fashion, some kind of institutional mechanism(s) need to be established.

INSECTION B

For example, in many parts of the contemporary world, the market system is used as the primarymeans of rationing scarce resources. To serve as a rationing devise, the rationing devise for resource allocation and distribution may be based on either (a) the market system, or (b) the central (government) distribution system.

In the market system, price which is determined by the market forces of demand and supply serves as the rationing instrument. However, rationing through government intervention uses various non-price factors as rationing instruments.

1.3 Operation of a Market-Oriented Economy

In many parts of the contemporary world, the market system is used as the primary meansof rationing scarce resources. The diagram below is designed to show that the operation of a market-oriented economy is composed of the following elements.

Economic entities (households and firms). Households are the final users of goods and services and the owners of resources. In a market economy, given the resource scarcity the primary is to find effective ways to address the material needs of consumers (households). At least in principle, consumers' well-being is the primary of a market-oriented economy. Although households are final users of goods and services, firms enter the economic process as transformers of basic resources (labour, capital and natural resources) into goods and services, and this is done on the basis of consumers' preferences (demand).

Markets. Markets represent an institutional arena in which exchanges (buying and selling) of final goods and services and factors of production (labour, capital and natural resources) take place. Traditionally, economists group markets into two broad categories, namely product and factor markets. The product market is where the exchange of final goods and services occurs. In this market, demand and supply provide information about households and firms, respectively. The factor market refers exclusively to the buyingand selling of basic resources, such as labour, capital and natural resources. In this submarket, demand imparts market information about firms and supplt provides information about households. That is, households are the supplier of labour, capital and natural resources, while firms are the buyers, and in turn use these items to produce final goods and services for the product market. Clearly, then, the roles played in the factor market by households and firms respectively are the reverse of their roles in the product market.

In both the product and the factor markets, information about resource scarcity is transmitted through prices. These prices are formed through the interactions of market demand and supply; and, under certain conditions, market prices can be used as reliable indicators of both present and future resource scarcities. Furthermore, using prices from the product market, economists customarily measure aggregate economic performance ofa given economy or a country by the total market value of all the goods and services produced for final use within a given period, usually a year. This is called gross domestic product (GDP) when the total market value of the final goods and services produced is attributable to factors of production (labour, capital and natural resources) originating exclusively from a given country (more on this in the next section).

Nonmarket public and private institutions. A market does not function in a vacuum;that is, for a market to operate efficiently, ownership rights need to be clearly definedand enforced. This requires the establishment of public agencies designed to articulateand enforce the rules and regulation by which ownership right are attained, relinquished (transferred) and enforced. In addition, competition in the market place is fostered through public intervention in some instances. The public and private entities (social institutions) that legislate the rules for assigning resource ownership rights and regulate the degree of competition in the marketplace are represented by the box at the center of figure 1.1. on one view, what flows from this box to households, firms and markets is not physical goods but information services. In general, the main function of these flows of information is to ensure that economic agents (households and firms) are playing by some socially predetermined rules of the game. In this regard, ideally, social institutions are perceived as being rather like a conductor of a symphony orchestra or a traffic director at a busy intersection.

Viewed this way, social institutions have important economic functions. However, they should not be assumed to be either perfect or costless (North 1995). When they are functioning well, the information communicated through them can distort market signals (prices) and in so doing, significantly affect the allocation of scarce resources.

1.4 The Concept of Resources

In broad terms, resources can be defined as anything that is directly or indirectly capable of satisfying human wants. Resources can be classified from two perspectives which are;

1. Traditional economics resource classification and,
2. Environmental economics resource classification

1. TRADITIONAL ECONOMICS CLASSIFICATION OF RESOURCES

Traditionally the economic notion of resources classifies resources into three broad categories: land, labour and capital

Land: This refers broadly to natural resources which are the stock of living and non-living materials found in the physical environment, and which have an identifiable potential use to human begins.

It should be noted that land is a very inadequate expression for what in a wider context amountsto the natural resources base. This is because it could be misunderstood asa place to build factories, cities and physical infrastructures like hospitals, schools, etc. By understanding land to mean all non-man-made natural resources, the idea of what is in, on and over the land is included becomes clearer. Thus, in Environmental Economics, it is better to say "natural resources" rather than simply "land". Agricultural land, deposits of ferrous and non-ferrous minerals, water, fisheries, and other aquatic life, wilderness and its multiple products are examples of material resources.

Labour: Labour encompasses the productive capacity of human physical and/or mental efforts, measured in terms of ability to work or produce goods and services. Entrepreneurship is often included under labour.

Capital: This refers to a class of resources that are man-made for the purpose of creating a more efficient production process. In other words, capital is the stock of produced items available not for direct consumption, but for further production process. Examples include all sorts of machineries.

ECONOMIC ASSUMPTIONS ABOUT RESOURCES

Assumption 1: Resources are Factors of Production

It is rare that basic resources of labour, capital and natural resources are used for their direct consumption without modifications. Hence resources are viewed in economics as a means to produce final goods and services that are capable of directly satisfying human wants. This is tosay that basic resources are a means to an end and not end in themselves.

Assumption 2: Resources have no Intrinsic Value

Related to assumption 1 above is the notion that the economic value of resources is strictly anthropocentric. This implies that the economic value of any resource is defined by human needs and nothing else. This idea treat human as preeminent as all other resources are deemed to exist for humans' economy and not for themselves. Non-economic values of resources are not considered. This has important consequences for the conservation of biodiversity.

Assumption 3: Only Scarce Resources are of Economic Concern

In economic analysis, each of the above resource categories is of economic concern to the extent that they are scarce i.e found in limited quantities and/or quantities. Any resources that are not limited in supply is not of economic concern.

Assumption 4: (See Back)

Assumption 5: Resources are Fungible

This implies that resources are substitutable. That is, on kind resources (such as machine) can be replaced by another (such as labour) in the production process; or one type of energy resources (e.g., petroleum) can be replaced by another form of energy (such as natural gas). Fungibility implies that no particular resource is considered to be absolutely essential for production of goods and services. It should be noted however that fungibility does not in any way suggest an escape from general problem of resource scarcity because there is the extent to which resources substitution can occur in production.

Assumption 4: Resources are used in Combination

This means that to produce a good or service require that various forms of resources are combined together to effect transformation into the required good or service. For instance, producing bricks for building requires a combination of sand, water, human skill, cement, block-making machine, shovel e.t.c in certain proportions.

Resource combination and substitutability can be depicted by the following table. For simplicity, we assume there are just two resources, labour and capital to produce a given level of output.

Quantity of Output	Quantity of Labour	Quantity of Capital

Resources can also be classified according to whether they are replenishable or not. Thus, we have the following categories.

Renewable Resources: resources are said to be renewable if they are replaced by naturalprocesses at a rate comparable or faster than their rate of consumption by humans. In other words, renewable resources have a natural rate of replenishment sufficient to augment the stock. Thus, renewable resources naturally regenerate over time e.g., fish, trees, wildlife, grazing lands. The environmental economic issues revolve round the consideration of the impact of a renewable resource use (extraction or harvest) on the rateof replenishment.

If too much is harvested, the rate of replenishment may not be sufficient to leave enough resources for the future. If too little is harvested, opportunities for gains are lost. The harvest decision involves a comparison of marginal benefit with marginal cost. If $MB_{harvest} > MC_{harvest}$, more harvest is justified, otherwise (If $MB_{harvest} < MC_{harvest}$) further harvest is not advised.

Non-Renewable Resources: these are resources for which there is no replenishment or the rate of growth is so slow as to be imperceptible in human life span. Thus, for non-renewable resource, the natural rate of replenishment is negligible in terms of augmenting the stock of the resource. Examples include oil, gas, uranium, aluminum e. t. c. The three stages of non-renewable resource use to consider in economics are exploration, development and extraction. The exploration, development or extraction decision also involves a comparison of the marginal benefit to marginal cost.

THE CONCEPTS OF DEMAND, SUPPLY AND MARKET EQUILIBRIUM

1.0 Introduction

In a market-oriented economy, the majority of price and output decisions are determined in themarket through the market forces of demand and supply. The concept of demand and supply remains the most fundamental in economic analysis and constitute the backbone of a market economy.

1.1 Demand

Demand is the quantity of a commodity buyer wish to purchase at each conceivable price.

1.12 Demand Schedule: is a table showing how much of a given commodity a buyer would be willing to buy at different prices. The table below illustrates a demand schedule

Price	Qty
5	35
4	45
3	60
2	80
1	110

1.13 The Demand Curve: is a graph illustrating how much of a given commodity buyers would buy at different prices. Demand curves are usually derived from a demand schedule. Using the demand schedule above, we can obtain a demand curve by plotting the values of prices(vertical axis) against the values of quantity bought (horizontal axis) to obtain the downward sloping demand curve.

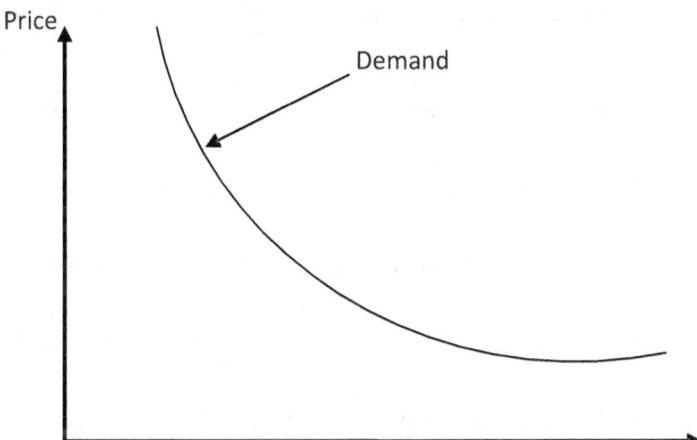

1.14 The Law of Demand: The law of demand states that other things being equal, as price increases, the quantity demanded of a commodity decreases. The law is saying that there is a negative or inverse relationship between price and quantity demanded of a commodity. This law is also illustrated by the downward sloping nature of the demand curve.

1.15 Explanation of the Law of Demand

Three reasons can be provided to explain the downward slopping nature of the demand curve.

a. **Diminishing Marginal Utility:** Utility is satisfaction derived from the consumption of a commodity. Marginal utility is the satisfaction derived from the consumption of one more extra unit of the commodity. The law of demand can be explained by diminishing marginal utility. The explanation is this: the decrease in added satisfaction as one consumes additional units causes one to consume more units in order to get the same level of satisfaction.

b. **Income Effect:** A lower price increases the purchasing power of money income enabling the consumer to buy more at lower prices without having to reduce demand for other goods and conversely.

c. **Substitution effect:** A lower price induces the consumer to buy more of the good with the lower price and less of the relatively high-priced substitutes.

1.16 Demand for a good or service can be defined for an individual household, or for a group of households that make up the demand side of the market. Market demand is the sum of individually demanded quantities at the various prices. The table below illustrates the derivation of a market demand schedule assuming there are two households or individual in the market.

Rice (A)	Quantity Demanded by Industry 1 (B)	Quantity Demanded by Industry 2 (C)	Market Demand (D)= (B) + (C)
3	40	60	100
4	30	40	70

The transition from individual demand curve to a market demand curve is similarly done as illustration below;

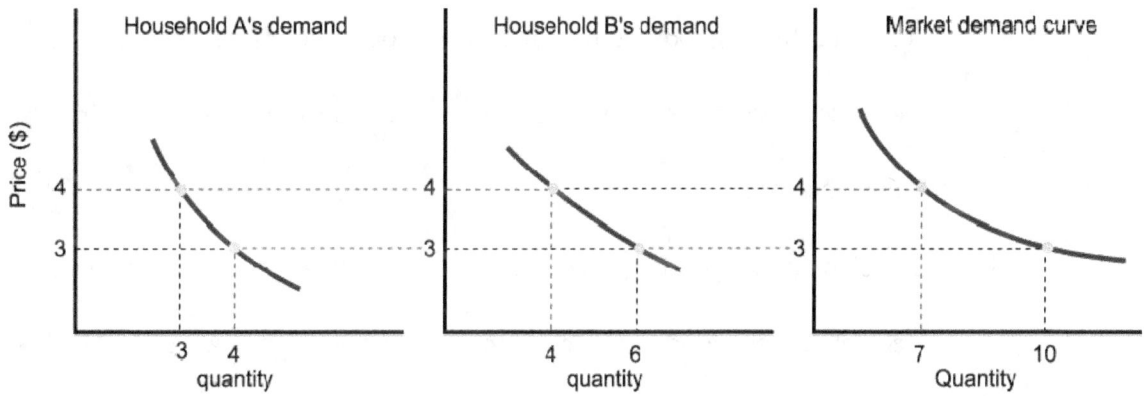

1.17 OTHER DETERMINANTS OF DEMEND

a. **Tastes** – favourable change leads to increase in demand; unfavourable change to decrease.

b. **Number of buyers** – more buyers lead to an increase in demand; fewer buyers lead to a decrease.

c. **Income** – more income led to an increase in demand; less leads to decrease in demand for normal goods. (The rare case of goods whose demand varies inversely with income is called inferior goods).

d. **Prices of related goods also affect demand**

   i. Substitute goods (those that can be used in place of each other): the price of the substitute good and demand for the other good are directly related. If the price of coke rises (because of a supply decrease), demand for Pepsi should increase.

   ii. Complementary goods (those that are used together like tennis ball and rackets). When goods are complements, there is an inverse relationship between the price of one and the demand for the other.

e. **Consumer expectations** – consumer view about future prices and income can shift demand.

1.18 CHANGE IN DEMAND

This means a shift of the entire demand curve to the right or to the left of the initial demand curve. A shift of the demand curve to the right indicates an increase in demand while a shift to the left implied a decrease in demand as a result of changes in the determinants of demand expect price. The figures below illustrate these two possibilities;

a. (i) Increase in Demand

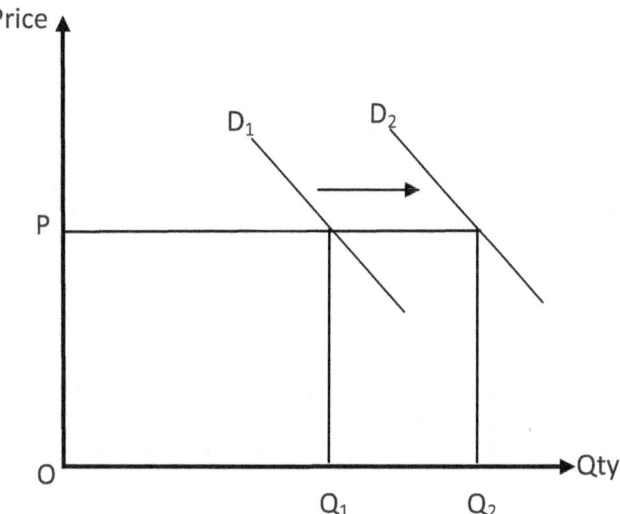

Notice that at the same price (p), a forward shift in demand from D1 to D2 cause the quantity demanded to increase from Q1 to Q2 even when price remains constant.

(ii) A summary of what can cause an increase in demand

    a) Favourable change in consumer tastes.

    b) Increase in the number of buyers.

    c) Rising income if product is a normal good.

    d) Falling incomes if product is an inferior good.

    e) Increase in the price of a substitute good.

    f) Decrease in the price of a complementary good.

    g) Consumers expect higher prices or incomes in the future.

a. (i) Decrease in demand

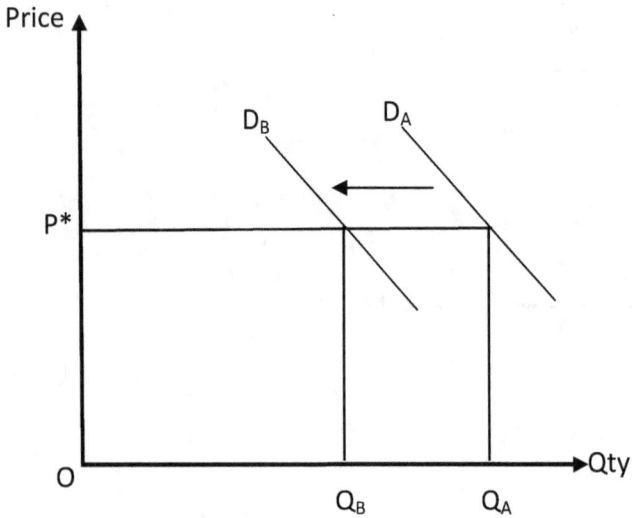

Notice that with decrease in demand, lower quantity that is demanded at the same price P*, asdemand shifts backwards from DA to DB.

(ii) **A summary of what can cause a decrease in demand.** This is presented below;

   a) Unfavourable change in consumer tastes.

   b) Decrease in the number of buyers.

   c) Falling income if product is a normal good.

   d) Rising incomes if product is an inferior good.

e) Decrease in price of a substitute good.

f) Increase in price of a complementary good.

g) Consumers' expectation of lower prices or incomes in the future.

1.19 Change in quantity demanded refers to movement along a demand curve as a result of changes in the price of the commodity.

A change in quantity demanded can be in two respects – **increase or decrease.**

i. Increase in quantity demanded occurs as a movement downwards on a demand curve when price falls and other determinants of demand remain constant.

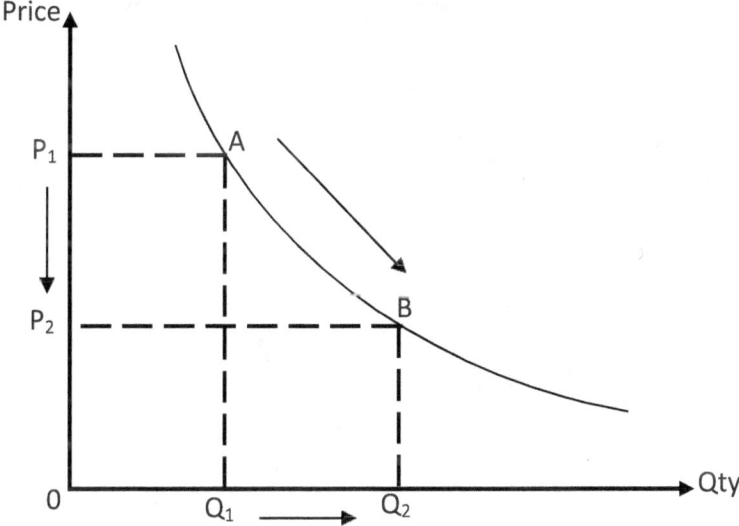

Increase in quantity demanded from Q1 to Q2 as price reduces from P1 to P2.

ii. Decrease in quantity demanded occurs as an upward movement along a demand curvewhen price of the commodity increases but other demand determinants remain constant.

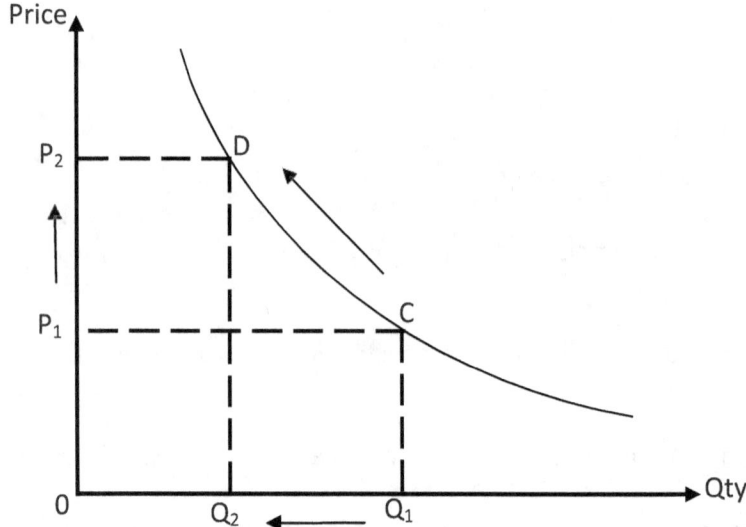

Decrease in quantity demanded from Q1 to Q2 when price rises from P2 to P1

1.2 SUPPLY

1.21 Supply is the quantity of a commodity sellers wish to sell at each conceivable price.

1.22 Supply Schedule is a table showing how much of a given commodity a seller would be willing to sell different prices. The table below illustrates a supply schedule.

Price	Quantity
2	15
4	20
6	25
8	30
10	38

Notice that as price increases from 2 to 10-dollars, Quantity demanded increase from 15 to 38units.

1.23   The Supply Curve is a graph illustrating how much of a given commodity sellers (firms) would sell at different prices.

Supply curves are derived from supply schedules by plotting price against the quantity supplied with price on the horizontal axis.

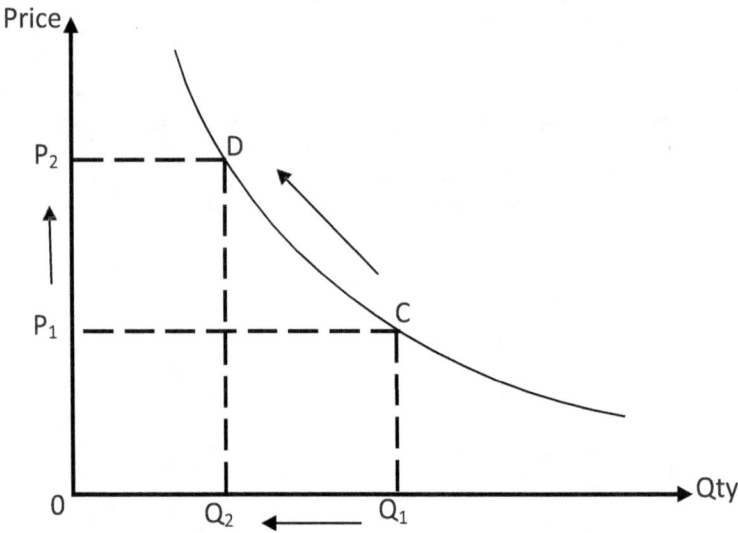

Notice that the supply curve is upward sloping and hence has a positive slope.

1.24   The law of supply states that the higher the price of a commodity. The higher thequantity supplied, if all other things are unchanged.

1.25   Explanation of the law of supply: with given product costs (i.e if cost of production remains unchanged), a higher price means greater profits and thus an incentive to increase thequantity supplied

1.26   **Market Supply**: the supply for a good or service can be defined for an individual firm, or for a group of firms that make up the supply side of the market.

Market Supply is the sum of all the quantities of a good or serviced per period by each by ach firm at the various prices.

As with market demand, *market supply* is the horizontal summation of individual firms' supply curves.

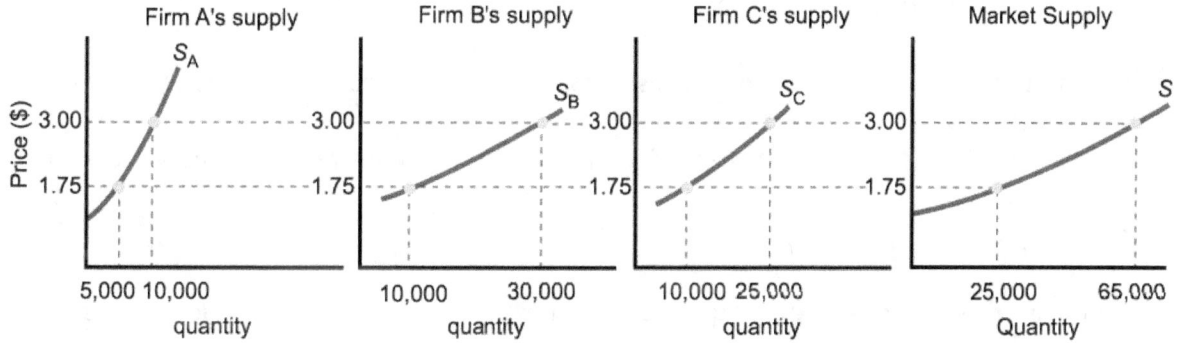

2.25 Other determinants of supply. These are factors (or variables) that determine supply except price of the commodity. These factors are briefly discussed below.

a) **Resource prices**- a rise in resource prices will cause a decrease in supply or leftward shift in supply curve; a decrease in resource prices will cause an increase in supply or rightward shift in the supply curve.

b) **Technology**- a technological improvement means more efficient production and lower costs, so an increase in supply, or rightward shift in the curve results.

c) **Taxes and subsidies**- a business tax is treated as a cost, so decrease supply; a subsidy lowers cost of production, so increases supply.

d) **Prices of related goods**- if price of substitute production good rises, producers might shift production toward the higher priced good, causing a decrease in supply of the original good.

e) **Producer expectations**- expectations about the future price of a product can cause producers to increase or decrease current supply.

f) **Number of sellers**- generally, the higher the number of sellers the grater the supply.

www.ingramcontent.com/pod-product-compliance
Lightning Source LLC
Chambersburg PA
CBHW082221220526
45470CB00010B/3259